THE MOUNTAINS

THIS EARTH OF OURS

Mel Higginson

The Rourke Corporation, Inc.
Vero Beach, Florida 32964

Edited by Sandra A. Robinson

PHOTO CREDITS
© Mel Higginson: all photos except page 17 © James P. Rowan
and page 18 © Frank Balthis

Library of Congress Cataloging-in-Publication Data

Higginson, Mel, 1942-
   The mountains / by Mel Higginson.
   p. cm. — (This earth of ours)
   Includes index.
   ISBN 0-86593-383-9
   1. Mountain ecology—Juvenile literature. 2. Mountains—
Juvenile literature. [1. Mountain ecology. 2. Ecology.] I. Title.
II. Series: Higginson, Mel, 1942-   This earth of ours.
QH541.5.M65H52· 1994
574.5'264—dc20   94-10721
                                             CIP
                                             AC

Printed in the USA

TABLE OF CONTENTS

Mountains   5
Where Mountains Are   6
Life in the Mountains   9
How Mountain Animals Live   11
A Mountain Animal: The Mountain Goat   14
People of the Mountains   16
How People Live on Mountains   19
The Mountain Community   20
A Mountain Community: The Alps   22
Glossary   23
Index   24

MOUNTAINS

Mountains are the world's high country — the slopes, peaks and meadows that tower over the land below.

Some mountains are great, snow-capped giants with rocky cliffs and sharp peaks. Others have rounded, forested tops.

Living high in the mountains is difficult. However, small groups of people and several kinds of plants and animals are at home there.

Mountains are the world's high country, where earth meets clouds

WHERE MOUNTAINS ARE

Mountains are great masses of rock and earth that rise on every continent. About one-fourth of the Earth's land surface is **mountainous.**

North America has several chains, or ranges, of mountain peaks. The Rocky Mountains of western North America make up the longest chain.

The world's most mountainous country is Tibet. Much of this Asian nation is more than three miles — 15,840 feet — above sea level!

*The Rockies are
North America's backbone*

LIFE IN THE MOUNTAINS

Life in the mountains is harsh, especially on tall mountain peaks. For every 1,000 feet of height above sea level, a mountain is about three degrees (Fahrenheit) cooler.

Wind, rain and snow are common on the upper slopes of mountains. Often this highest land is treeless. Trees cannot grow on the rocky, windswept ground.

Even so, mountains are **habitats,** or homes, for many plants and animals. Golden eagles, mountain lions, **marmots** and mountain goats are just a few of the animals living in the mountains of western North America.

A Dall sheep stands on a snowy mountain ledge in the Yukon Territory of Canada

HOW MOUNTAIN ANIMALS LIVE

Mountain animals have special ways to live in the high country. Mountain sheep and goats have special hoofs to grip the rocks. These animals can run easily up and down steep, rocky slopes.

Most animals leave the mountaintops as winter comes. A few can withstand mountain winters, however. Marmots **hibernate,** or sleep, winter away in deep burrows.

A marmot snacks on mountain greens in early September, shortly before hibernating

The snow leopard is a rare predator of Himalayan peaks in Asia

The amazing yak is a walking supermarket for people in the Himalayas

A MOUNTAIN ANIMAL: THE MOUNTAIN GOAT

Mountain goats spend all year high in the mountains of western North America. Mountain goats survive the winter cold by having two coats of hair.

The lack of winter food is a problem for most mountain animals, but not for mountain goats. Mountain goats need less food than other animals their size. They survive on bits of plants their hoofs uncover from beneath the snow.

Unlike most animals of the heights, mountain goats never leave the high country

PEOPLE OF THE MOUNTAINS

People live in mountain villages throughout the world. Some villages are quite modern. Others, scattered in South America, Africa and Asia, are **primitive** — they lack modern things. In primitive villages, people live in the old ways.

Mountain life can be hard. Steep mountainsides make travel difficult. When people are active, even breathing is more difficult than it is at sea level. Mountain air is "thinner" than air below. That means it has less oxygen, which we need for breathing.

Villagers gather to buy and sell crafts and food products at an outdoor mountain market in Ecuador

HOW PEOPLE LIVE ON MOUNTAINS

Mountain people who live in the old ways depend upon little farms and herds of goats, sheep or cattle for their food and clothing. They cut trees out of mountain forests for firewood.

In the Himalayan Mountains of Asia, villagers raise the amazing, cowlike yak. The yak hauls wagons and gives milk. It provides meat, leather and wool. Herdsmen use its **dung,** or waste, for fuel and fertilizer.

Mountain farmers in Ecuador herd
sheep and burros along a lane

THE MOUNTAIN COMMUNITY

The plants and animals of the mountains are important to each other. Together they form a natural community.

With their roots, plants keep mountain soil from washing away. They also provide food for many mountain animals. Insects, goats, sheep and even grizzly bears eat plants.

Mountain **predators** live by eating plant-eaters. The mountain lion, for example, hunts deer.

A Swiss village lies nestled in a green valley in the Alps

A MOUNTAIN COMMUNITY: THE ALPS

The Alps of Europe are mountains with jagged, rocky peaks and deep, green valleys. Little villages are scattered throughout the valleys. Some of the Alpine people live above the villages in homes called **chalets.**

Like mountain farmers elsewhere, Alpine farmers — French, Swiss and Italian — have a short growing season. They raise mostly goats, sheep and cattle, and hay and clover for their herds.

Wildflowers brighten Alpine meadows. Marmots, **ibex** and **chamois** graze on Alpine plants.

Glossary

chalet (sha LAY) — a farmer's hut or home in the Alps

chamois (SHA mee) — a goatlike antelope of the Alps

dung (DUNG) — the solid waste of an animal

habitat (HAB uh tat) — the special kind of area where an animal lives, such as *mountain forest*

hibernate (hi ber NAYT) — to enter a long, deep winter sleep during which an animal's normal body functions are slowed

ibex (I beks) — a wild goat with long, backward-curving horns

marmot (MAR mut) — a large, burrowing ground squirrel of the high mountains

mountainous (MOUN tin us) — relating to mountains

predator (PRED uh tor) — an animal that kills other animals for food

primitive (PRIM uh tihv) — relating to the earliest ways of doing things; not modern

INDEX

Africa   16
air   16
Alps   22
animals   5, 9, 11, 14, 20
Asia   6, 16, 19
cattle   19, 22
cliffs   5
Europe   22
farmers   22
farms   19
goats   19, 20, 22
herds   19, 22
Himalayan Mountains   19
hoofs   11, 14
marmot   9, 11, 22
meadows   5
mountain goat   9, 11, 14
mountain lion   9, 20
North America   6, 9, 14

peaks   5, 6, 9, 22
people   5, 16, 19
plants   5, 9, 20, 22
rain   9
Rocky Mountains   6
sheep   19, 22
    mountain   11
slopes   5, 9
snow   9, 14
soil   20
South America   16
Tibet   6
trees   9, 19
valleys   22
villages   16, 22
wind   9
winter   11
yak   19